The journey of the stem cell

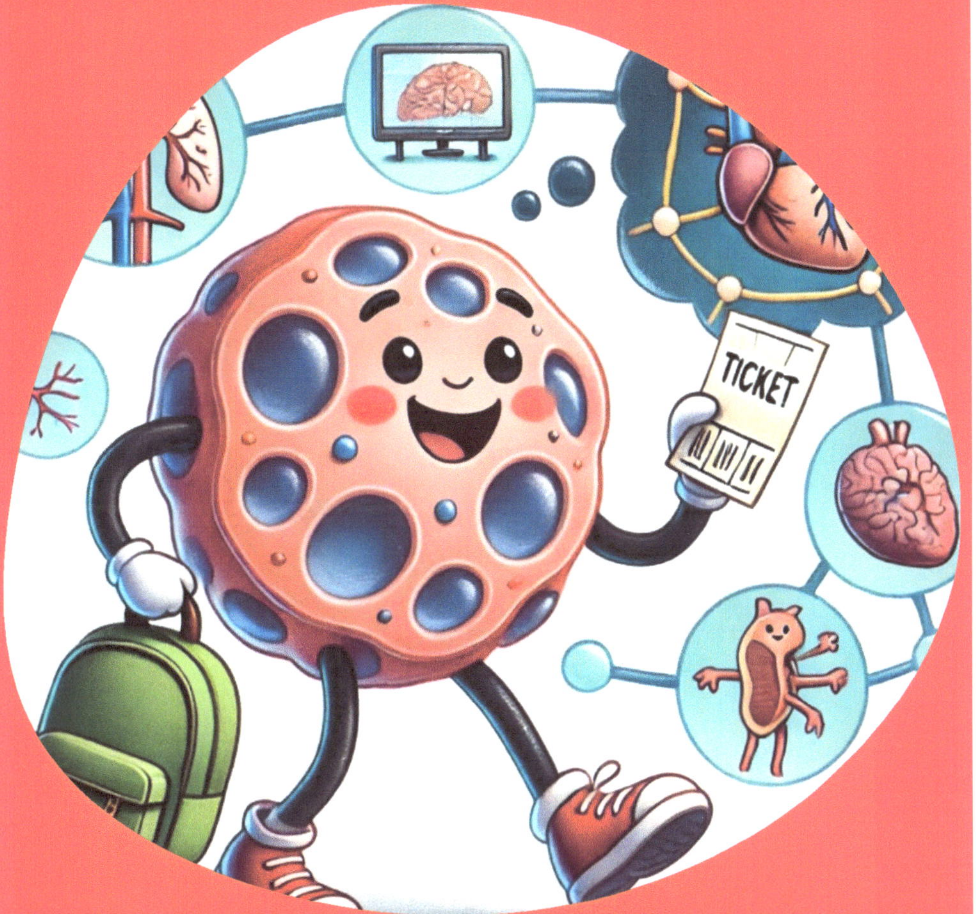

Jorge Luis Maria Ruiz

The journey of the stem cell

Jorge Luis Maria Ruiz

Dedication

This book is dedicated to my wife, Roberta,
and my daughter, Helena.
And most importantly, to God, through whom all things are possible.

Preface

This work was created with the aim of introducing children to the fascinating world of biology in a simple and engaging way through an adventure. The Journey of the Stem Cell offers a first glimpse into the wonderful world of cellular biology, presenting the birth and role of a stem cell in the human body.

Stem cells have the incredible ability to regenerate different tissues, thanks to the process of cellular differentiation, which can be stimulated in various ways. In the story, our little stem cell doesn't yet know what type of cell it wants to become. To make this important decision, it embarks on a journey through the body, talking to other adult cells.

During this journey, the reader will have the opportunity to discover, in a playful manner, the main functions of each cell encountered. By the end of this long and enriching journey, the stem cell will have the knowledge needed to choose its destiny.

Summary

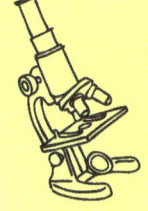

Chapter 1: The Birth of the Little Cell ... 07

Chapter 2: The Meeting with the Red Cell ... 12

Chapter 3: The Meeting with the White Cell ... 18

Chapter 4: The Meeting with the Muscle Cell ... 24

Chapter 5: The Meeting with the Brain Cell ... 30

Chapter 6: The Decision of the Little Cell ... 39

Glossary ... 49

Chapter 1: The Birth of the Little Cell

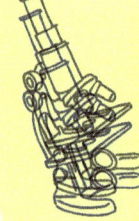

Once upon a time, inside a very special body, there was a little cell that had just been born. It came about through something called cell division - a magical process that turns one cell into two! But this was no ordinary cell. It was a stem cell!

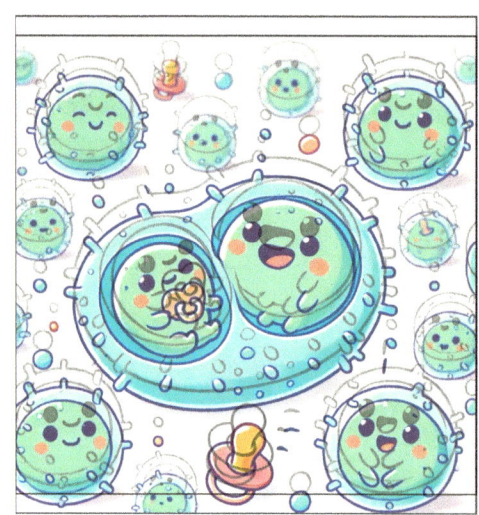

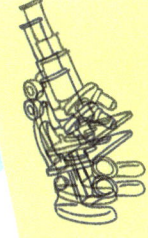

Only, unlike other cells, the little stem cell didn't yet know what it was going to be when it grew up. It had no idea what its function in the body would be or which organ it would work in. And that made her a little worried.

- she wondered, full of curiosity.

"Will I become a heart cell? Or maybe a skin cell? Or maybe a muscle cell?"

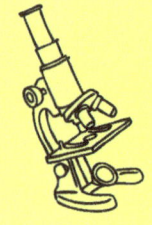

As the little cell pondered its future, an older stem cell appeared. She was already experienced and knew everything about the human body. With a smile, the older stem cell said:

"You still have plenty of time to choose what you're going to be. Why don't you take advantage of it and get to know the human body? Talk to other cells and find out which function makes you happiest!"

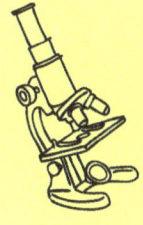

The little cell, excited by the idea, replied:

"I'd love to do that! Where should I go first?"

The older stem cell winked:

"Follow the bloodstream! It will take you to many amazing places and introduce you to different types of cells."

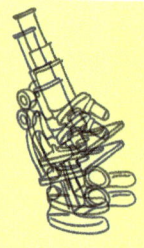

And so, full of excitement and curiosity, the little stem cell embarked on its first great adventure inside the body. Moved by the flow of blood, she was ready to meet new friends and discover what the future had in store for her!

Chapter 2: Meeting the Red Cell

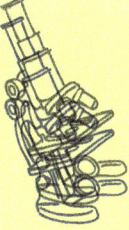

The little stem cell traveled happily through her bloodstream. Her body was huge, and she couldn't wait to find out who she would meet along the way. Then, all of a sudden, a little red cell came gliding up beside her.

"Hi, who are you?"

- asked the stem cell, curious.

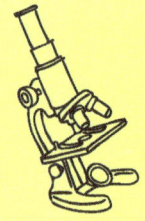

"I'm a red blood cell, but you can call me a red cell!"

- replied her new friend, smiling.

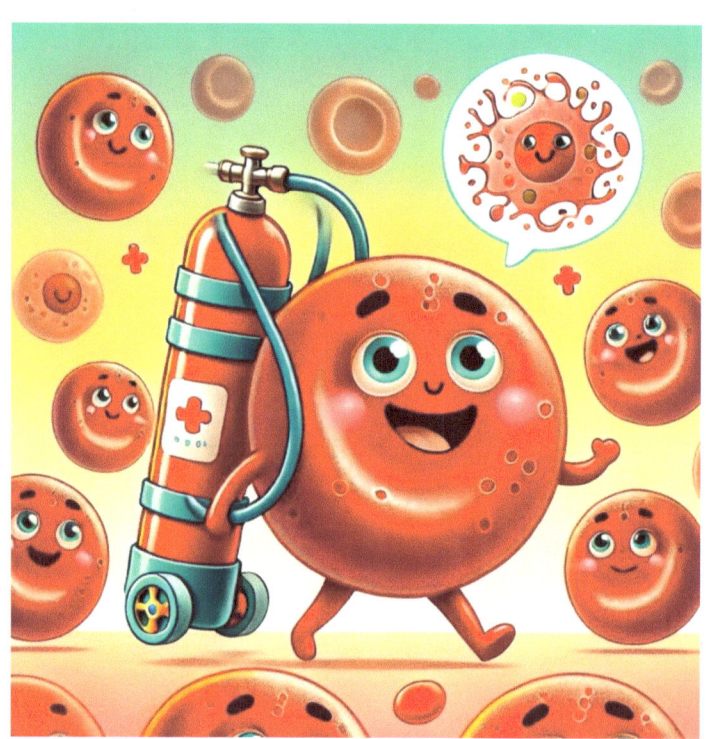

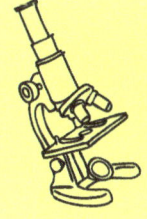

"That's great! And what do you do here in your body?"

- asked the stem cell.

The red cell puffed out its chest proudly and said:

"I have a very important job! I'm responsible for transporting oxygen. You know that fresh air you breathe into your lungs? Well, I take that oxygen to all the tissues in your body, even to your big toe!"

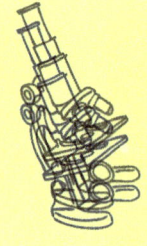

The stem cell was impressed.

"Wow, what a responsibility! Without you, the body wouldn't be able to breathe properly, would it?"

"Exactly! Without oxygen, the other cells wouldn't be able to function properly. That's why my job is so important. I make many journeys, going from the lungs to all parts of the body and then back to get more oxygen."

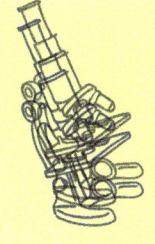

The little stem cell turned pensive.

"Being a red cell sounds amazing! Could I become one of you someday?"

"Absolutely, if you want to!" - said the red blood cell. "But the body needs many different types of cells to function well. You'll meet other cells with super interesting jobs!"

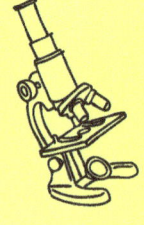

The little stem cell smiled and said goodbye to the red cell.

"Thanks for telling me about your work! I'm going to continue my journey. Who knows what else I'll discover?"

And with that, the stem cell moved on, floating through the bloodstream and eager to learn more about the human body.

Chapter 3: Meeting the White Cell

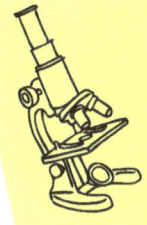

The little stem cell was making its way through the bloodstream when it suddenly spotted a different cell. It was bigger, white, and seemed to be always on the alert, as if ready to face anything.

"Hello, who are you?"

- asked the stem cell, coming closer.

"I'm a lymphocyte, but you can call me a white cell!"

- replied the cell, with a determined look on its face.

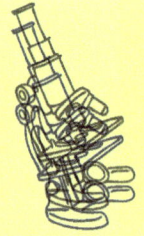

"That's nice! And what do you do here in the body?"

- asked the stem cell, curious.

The white cell smiled and explained:

"I'm part of the body's defense system! My main function is to protect the body from invaders such as bacteria and viruses. When something dangerous enters the body, like a nasty microbe, I and my white cell friends are the first to spring into action."

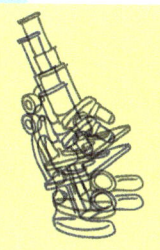

"Wow!"

-- exclaimed the stem cell, impressed:

"So you're like a body guard, protecting the body against disease?"

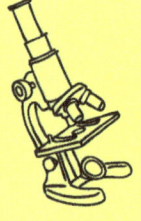

"Exactly!"

"We fight these threats and also produce antibodies, which are like special little weapons that help destroy these invaders. Thanks to us, the body is able to stay healthy and recover when it gets sick."

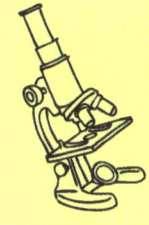

The stem cell marveled at the white cell's courage.

"It must be very exciting to protect the body! Have you ever had to fight off many invaders?"

"Oh, several times!"

- replied the lymphocyte.

"And we're always ready to act, because the body needs a strong defense. Without white cells, the body could easily get sick."

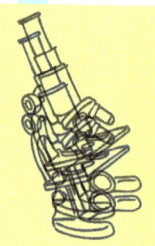

"That's amazing!" - said the stem cell in amazement.
"Who knows, maybe I too will become a white cell one day..."

"Maybe, but remember, the body needs many types of cells to function well. Continue your journey and you'll see that each one has a special function!"

The little stem cell thanked her and moved on, more excited than ever to discover other cells and their incredible functions.

Chapter 4: Meeting the Muscle Cell

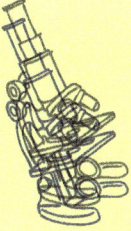

The little stem cell continued its adventure, now leaving the bloodstream to explore other parts of the body. Suddenly, it arrived in a different place: the muscles! There, everything seemed strong and resilient. It was then that she found a very long cell full of energy.

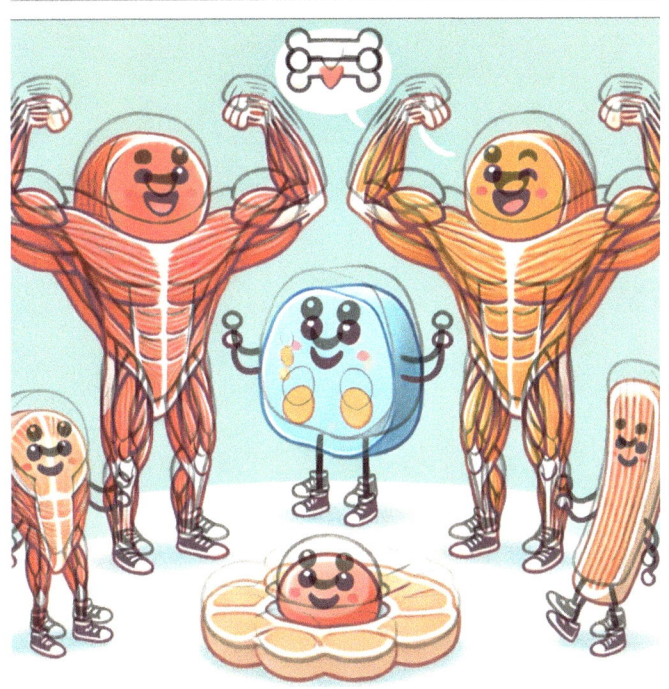

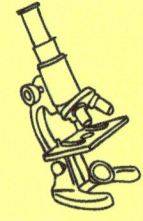

"Hi, who are you?"

- asked the stem cell.

"I'm a muscle cell!"

- replied the new friend, as it stretched and contracted with ease.

"We muscle cells have a very important job. We're the ones who give the body strength and support!"

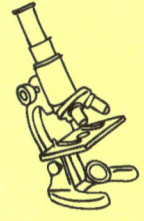

The stem cell was curious.

"What do you mean you give strength?"

The muscle cell smiled and said:

"Well, we form the muscles that are all over the body, such as those in the arm, leg and even the heart! When you lift a toy or run in the park, it's the muscles that are working. They contract and relax all the time, allowing the body to move and be strong."

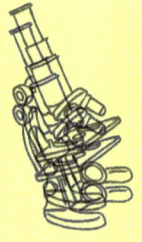

"Wow, so you're the one who helps the body stay strong and moving?"

- asked the stem cell, amazed.

"That's right!"
"We also work non-stop in the heart, helping to pump blood throughout the body. Without us, the body wouldn't have the strength to move or even to keep the heart beating!"

- replied the muscle cell.

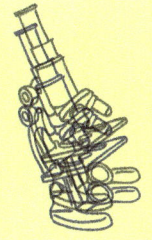

The stem cell thought for a moment.

"Wow, I'd never thought about how important muscles are to all this! It must be very rewarding to help the body get so strong."

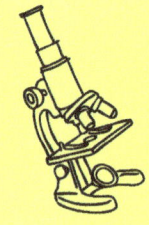

The muscle cell smiled and said,

"It sure is! But, as you may have noticed, each cell plays a fundamental role. Without the muscles, the body couldn't move, but without the blood and the other cells, it wouldn't function properly either. It's teamwork that makes everything work!"

The little stem cell was grateful for the explanation and once again continued its journey, increasingly fascinated by the different functions of cells.

Chapter 5: Meeting the Brain Cell

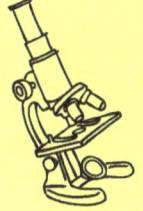

After meeting the muscles, the little stem cell was excited to continue its journey. This time, it found itself in a very special place: the brain! There, everything seemed to be in constant activity, with little electrical signals running back and forth like tiny rays. That's when she found a different cell, which seemed to be in charge of everything.

"Hi, who are you?"

- asked the stem cell, enchanted by the movement around it.

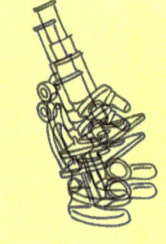

"I'm a neuron, a brain cell!"

- replied the new friend, as its long arms, called dendrites and axons, stretched out in all directions.

The stem cell opened its eyes wide with curiosity.

"Wow! What are you doing here in the body?"

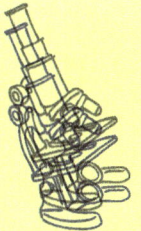

The neuron smiled and said:

"My job is very important! I help control and organize everything that happens in the body. I'm responsible for sending quick messages between the brain and the rest of the body. When you think, move, smell, hear sounds or even remember something, it's me and my neuron friends who make it happen!"

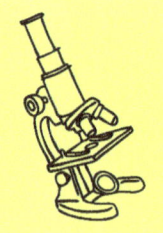

The stem cell was impressed.

"So, you're like the boss who controls all the other cells?"

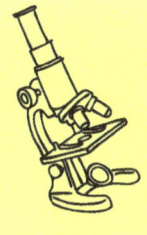

"In a way, yes!"

- replied the neuron.

"We neurons work together to make sure that the body knows what's going on around it and inside it. And we don't do it alone. The brain is like a control center full of neurons, and each of us has a specific function. Together, we send electrical signals that travel through the spinal cord and nerves so that the body functions perfectly."

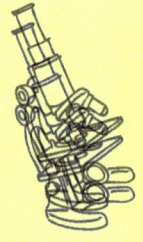

"That's incredible!"

- said the stem cell in amazement.

"You really do sound like the great organizers of the body."

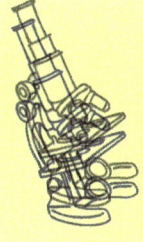

The neuron agreed with a nod.

"It's a complex job, but very important. Without us, the body wouldn't know how to react to things. For example, if you touch something hot, our signals tell the brain to move your hand before you burn yourself. All this happens in a matter of seconds!"

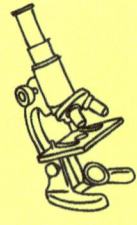

The little stem cell was delighted with everything it learned.

"I'd love to be a neuron! Controlling the body sounds like a big responsibility."

"It sure is!"

- said the neuron.

"But remember, all cells are equally important in keeping the body functioning."

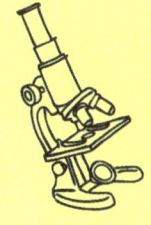

After saying goodbye to the neuron, the stem cell decided that it had already learned a lot, but it wasn't ready to choose yet. It knew that the body was vast and full of surprises, and it wanted to keep discovering more.

Now it was ready to go back and talk to the adult stem cell.

Chapter 6: The Little Cell's Decision

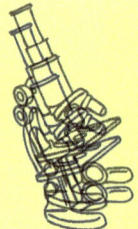

After a long and exciting journey, the little stem cell returned to its starting point. She felt different now - full of new ideas and her heart pounding from so much learning. It was then that it met the adult stem cell, which was waiting for it with a warm smile.

"Welcome back! How was your journey through your body?"

- asked the adult stem cell.

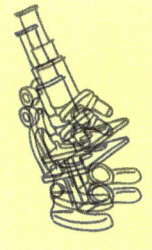

The little cell excitedly began:

"It was amazing! I met so many different cells! First, I met a red cell, which carries oxygen to all the body's tissues. Then I met a white cell, which defends the body against invaders. I also met a muscle cell, which gives the body strength and support, and finally a neuron, which controls and organizes everything in the brain."

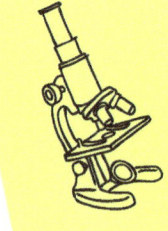

The adult stem cell smiled with satisfaction.

"Looks like you've learned a lot! So, have you decided what you want to be when you grow up? Which cell would you like to become?"

The little stem cell stopped for a moment and looked around. She had liked all the cells she had met, and each one seemed to have such an important and special role in the functioning of the body. But there were so many other cells to meet, so many other places to explore!

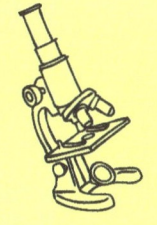

"I liked all the cells I met!"

- replied the little cell.

"But I don't think I'm ready to choose yet. I want to keep exploring and learning more about the body. There are still so many amazing functions that I didn't know about! I think I'll wait a little longer before deciding."

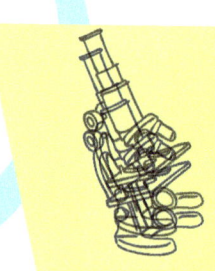

The adult stem cell laughed gently and said:

"That's a great decision! You have time, and the most important thing is to get to know your body well before choosing what makes you happiest. Continue your journey, and when the time is right, you'll know which way to go."

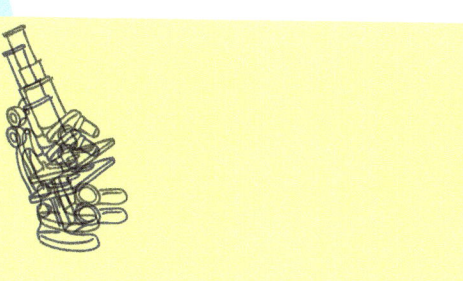

The little stem cell smiled, relieved and happy to know that there was no need to rush. It was ready to continue its adventure through the human body, getting to know more cells, learning about their functions and, who knows, one day discovering which one it wanted to be.

And so our little stem cell moved on, full of dreams and curiosity, exploring the incredible world inside the human body.

The end

Glossary

1. Stem Cell

Stem cells are special cells that have the ability to transform into various other types of cells in the body, such as muscle, nerve or blood cells. They can also divide many times, creating new cells, which helps with growth, repairing damaged tissue and replacing old cells. The main characteristic of stem cells is their ability to regenerate and differentiate, which means that they can "decide" which type of cell they will become to help the body function properly.

2. Cell division

Cell division is the process by which a cell divides into two new cells. This process is fundamental to the growth, reproduction and repair of the body's tissues. There are two main types of cell division:

- Mitosis: where a cell divides to form two identical cells, which contain the same genetic material. This is common in growth and regeneration processes.
- Meiosis: occurs for the formation of reproductive cells (such as sperm and eggs) and results in cells with half the amount of genetic material.

The main characteristic of cell division is the ability to create new cells, ensuring the correct functioning of the body and the continuity of life.

3. RBC (Red Blood Cell)

RBCs, also known as red blood cells, are blood cells responsible for transporting oxygen from the lungs to all parts of the body and removing carbon dioxide from the tissues, taking it back to the lungs to be eliminated. They contain a protein called hemoglobin, which binds to oxygen, giving the blood its characteristic red color.

Glossary

4. Lymphocytes (White Defense Cells)

Lymphocytes are a type of white blood cell whose main function is to defend the body against infections and invading agents such as viruses and bacteria. They are part of the immune system and help fight disease in two main ways:
- T lymphocytes: directly attack infected or cancerous cells.
- B lymphocytes: produce antibodies, which are proteins that help neutralize invaders.

5. Muscle cell

Muscle cells are specialized cells that form the body's muscles. They have the main function of conducting movements and providing support to bones and other structures. There are three main types of muscle cells:
- Skeletal Muscle Cells: These form the muscles that move the bones and are responsible for voluntary movements, such as walking and lifting objects.
- Cardiac Muscle Cells: Found in the heart, they are responsible for contracting the heart and pumping blood throughout the body.
- Smooth Muscle Cells: Located in the walls of internal organs such as the stomach and intestines, they control involuntary movements such as peristalsis.

Glossary

6. Neuron

Neurons are specialized cells in the nervous system responsible for transmitting and processing information through electrical and chemical signals. They form the basis of the brain, spinal cord and nerves. Each neuron is made up of three main parts:

- Dendrites: Receive signals from other neurons and send them to the cell body.
- Cell Body: Contains the nucleus and other structures essential to the neuron's function.
- Axon: Transmits electrical signals from the cell body to other neurons, muscles or glands.

Biography of the author

The author is Helena's father and an Associate Professor at the Federal University for Latin American Integration (UNILA). He holds a degree in Genetics Education from the National University of Misiones (Argentina, 2006) and a Bachelor's degree in Biological Sciences from UNESP (2009). He earned a Ph.D. in Sciences from the University of São Paulo's School of Medicine (2011) and completed postdoctoral research at Cold Spring Harbor Laboratory (2013). Currently, he teaches undergraduate courses in Biotechnology and Biological Sciences, as well as a graduate program in Biosciences at UNILA. He coordinates the Applied Health Biotechnology Laboratory at UNILA. His research focuses on Cellular and Molecular Biology, with an emphasis on Genetics, working mainly in the areas of stem cells, regenerative therapy, gene therapy, cellular and molecular biology, pharmacology, and cancer.

This work was created with the aim of introducing children to the fascinating world of biology in a simple and engaging way through an adventure. The Journey of the Stem Cell offers a first glimpse into the wonderful world of cellular biology, presenting the birth and role of a stem cell in the human body.

In the story, our little stem cell doesn't yet know what type of cell it wants to become. To make this important decision, it embarks on a journey through the body, talking to other adult cells.

During this journey, the reader will have the opportunity to discover, in a playful manner, the main functions of each cell encountered. By the end of this long and enriching journey, the stem cell will have the knowledge needed to choose its destiny.